Cover illustration: Short SC.1 in vertical flight. Note the open area in the fuselage undersurface to permit operation of the lift-jets.

1. In 1950 the US Navy held a design competition for a sea-going VTOL fighter that could be flown from ships with small aircraft-landing decks. Lockheed and Convair received contracts to produce flying prototypes, each of which was powered by one Allison YT40-A turboprop engine, driving co-axial contra-rotating propellers. Known as tail-sitters, because of their designed vertical launch position, the first to fly was the Lockheed XFV-1 Vertical Riser, in March 1954. However, this never flew vertically, having been initially fitted with a temporary fixed landing gear, as seen.

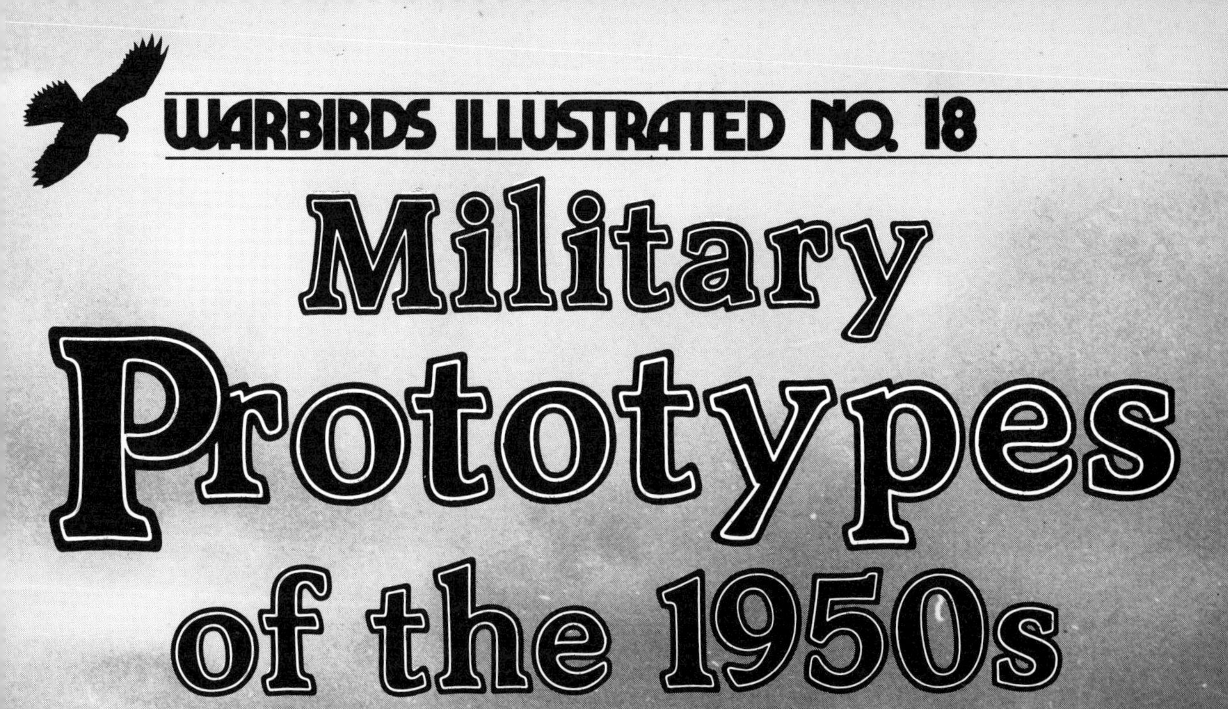

WARBIRDS ILLUSTRATED NO. 18

Military Prototypes of the 1950s

MICHAEL J. H. TAYLOR

a&ap
ARMS AND ARMOUR PRESS
London—Melbourne—Harrisburg, Pa.

U.S. ARMY
HILLER
U.S. ARMY

Introduction

Warbirds 18: Military Prototypes of the 1950s
Published in 1983 by
Arms and Armour Press, Lionel Leventhal Limited, 2-6 Hampstead High Street, London NW3 1QQ; 4-12 Tattersalls Lane, Melbourne, Victoria 3000, Australia; Cameron and Kelker Streets, PO Box 1831, Harrisburg, Pennsylvania 17105, USA.

British Library Cataloguing in Publication Data:
Taylor, Michael J. H.
Military prototypes of the 1950s.—(Warbirds; 18)
1. Airplanes, Military—History
I. Title II. Series
623.74′6 TL685.3
ISBN 0-85368-579-7

Layout by Anthony A. Evans.
Printed in Great Britain by William Clowes (Beccles) Limited.

2. The Hiller 'Flying Platform', illustrated in its original form. Ducted propeller research by Hiller began in 1954 and development was so rapid that a first tethered flight by the platform was made possible on 21 January 1955. A free flight followed eleven days later. The development of the platform was directed by the Office of Naval Research and the US Army.

This volume is the first in the Warbirds Series to be devoted to non-production aircraft. It covers the post-Second World War decade most prolific in new military prototypes, as designers grappled with turbojet and turboprop engines to produce viable supersonic fighters and bombers, naval attack and VTOL (Vertical Take Off and Landing) aircraft, jet-powered flying-boats and a gamut of other types. Some proved to be the progenators of later production aircraft, but others disappeared from the aviation scene as quickly (but often more quietly) as they had arrived. It is on the prototypes that failed to spawn series-built aircraft that this book is based, although it includes a few military experimental aircraft built for research purposes and a number of prototypes from which followed extensively redesigned service types.

Where possible, the aircraft covered have been arranged in chronological order, with first flight dates as the points of reference. The first few aircraft made their maiden flights in the latter 1940s but are included here because they continued test flying into the 1950s. By adopting this order, the reader can discern advances in design as the 1950s progressed and technical skills were learned. For example, although it is remembered that Captain Charles 'Chuck' Yeager was the first man to pilot an aeroplane faster than the speed of sound, in 1947, his aircraft was rocket powered and this type of power plant proved generally unsuccessful as a source of primary thrust for possible series-built aircraft. The development of supersonic aircraft suitable for production using turbojet power proved much more difficult, giving rise also to a number of prototypes with 'mixed' rocket and turbojet power.

Michael J. H. Taylor, 1983

▲3 ▲4 ▼5

3. The awesome capabilities of long-range bombers and fighter-bombers in production towards the end of the Second World War gave rise to the post-war development of fighters that could operate independently of airfields. One such fighter was the British Saunders-Roe SR.A/1, the world's first turbojet-powered flying-boat. First flown in 1947 and armed with four 20mm cannon, the SR.A/1 lacked manoeuvrability and was abandoned. However, one prototype continued test flying and research work into the 1950s.
4. Three SR.A/1 prototypes had been ordered by the Air Ministry, the first bearing the serial TG263.
5. The third SR.A/1 prototype (TG271). Note the new fully-transparent cockpit canopy.
6. On 18 September 1948, the Convair Model 7002 successfully performed its first flight. The world's first turbojet-powered delta-winged aircraft, it had been built to test this wing configuration as part of the development programme for the XF-92 turbojet and rocket-powered fighter, and to prove the superiority of delta wings over sweptback wings.
7. After the cancellation of the XF-92 project, the Model 7002 was officially designated XF-92A. Capable of speeds of up to Mach 0.95, the XF-92A was transferred in 1952 from the USAF to the National Advisory Committee for Aeronautics (NACA), with whom it continued flying.

6▲ 7▼

▼8 ▲9

8. The first French official military programme to be undertaken after the end of hostilities in 1945 involved the development of the Sud-Ouest Espadon, a single-seat fighter to be powered by a French-built Nene turbojet engine. The first prototype was the 6020-01, which first flew in November 1948. The 6021 (as illustrated) was eventually developed from the next prototype. A lighter experimental fighter built in 1950, the 6021 was armed with 20m or 30mm cannon. Its maximum speed was about 1,000 km/hr (620mph).

9. The 6025 Espadon, first flown in December 1949, was the third prototype in modified form. The newly-fitted auxiliary SEPR 251 bi-fuel rocket motor and jettisonable fuel tanks can be seen under the fuselage.

10. The final aircraft in the Espadon series was the 6026, first flown on 15 October 1951. In fact the second prototype modified, it featured a rocket motor in the rear fuselage, positioned under the jetpipe for the turbojet engine.

10

▲11

11, 12. The British Armstrong Whitworth A.W.55 Apollo was a 'red herring' military prototype. Designed to the recommendations of the Brabazon Committee for post-war British airliners, two Apollo prototypes were ordered by the Ministry of Supply for testing, each originally carrying military serials and markings. Powered by four Armstrong Siddeley Mamba turboprop engines and accommodating 26-41 passengers, the first Apollo flew initially on 10 April 1949. The airliner received its Certificate of Airworthiness in 1950. No further examples were built.

13. Illustrated is the first of two completed prototype Supermarine ASR.I Seagull amphibious flying-boats. The Seagulls were ordered by the Air Ministry for evaluation as possible forerunners of similar general utility aircraft to operate from Royal Navy aircraft carriers. Deck landing trials on HMS *Illustrious* were conducted in October 1949. Before being scrapped in 1952, one set a world speed record for amphibians in a 100km closed circuit.

14. On 4 June 1949, Lockheed flew its new Model 90 experimental long-range penetration fighter, which received the USAF designation XF-90. A very advanced design, it was powered by two Westinghouse turbojets, featured wings with 35° of sweepback, and the pilot sat high in a cockpit affording excellent visibility. The design progressed no further.

▼12

13▲ 14▼

▲15 ▼16

15. 4 September 1949 was the day that the first British jet-powered delta-winged aircraft flew as the Avro Type 707. It had been designed to flight-test delta wings at low flying speeds, the data helping in the design of the Vulcan V-bomber. It was subsequently lost in an accident.
16. The Avro Type 707B (VX790) succeeded the 707 for low-speed delta research, first flying in September 1950. Note the unusual position of the air intakes for the single Rolls-Royce Derwent turbojet engine.

17▲ 18▼

17. On 14 June 1951, a third Type 707 flew as Type 707A (WD280), this time for high-speed research. This aircraft featured wing-root air intakes. Here, WD280 is seen alongside VX790.
18. In February 1953 the second Type 707A flew, followed on 1 July by the side-by-side two-seat Type 707C. The two-seater was used to give flying experience to future pilots of delta winged aircraft and outlasted the others in the series by many years. Interestingly, the first Vulcans had straight leading-edges to their wings in typical Type 707 fashion.

▲19 ▼20

▼21

19. Installed with its intended Armstrong Siddeley Mamba turboprop engine, driving co-axial contra-rotating propellers, the actual Y.B.1 prototype first flew on 19 July 1950. Here, the Y.B.1 (WB797) flies with the second Griffon-engined prototype on 28 July, the latter designated Y.A.8.

20. Before the Fairey Gannet was selected for FAA service as an anti-submarine aircraft, Fairey had a rival in the Blackburn Y.B.1. A three-seater with an internal weapons bay, folding wings and a retractable search radar, the first Blackburn prototype flew on 20 September 1949. But, as it had been fitted with a Griffon piston engine to speed up flight trials, this prototype was known as the Y.A.7 (WB781).

21. On 28 October 1949, at Baltimore, the first prototype of the Martin XB-51 attack bomber completed its maiden flight. This view of an XB-51 shows clearly the pilot's cockpit and the upper window to the navigator's compartment to its rear, the 35° of sweepback on the wings, and the air intake for the rear turbojet engine forward of the tail.

▲22

22. Powered by three General Electric J-47 engines, two mounted on the forward fuselage and one in the tail, the XB-51 failed to gain production orders. The ribbon-type braking parachute is seen deployed in this photograph.
23. Early in 1950 a second XB-51 began flying. Both aircraft were handed over to the USAF for service evaluation by 1952.
24. The US Republic XF-91 was an experimental mixed-power interceptor, carrying a General Electric J-47 turbojet engine and a Reaction Motors rocket motor. An unusual feature was its variable-incidence, inverse-tapered, sweptback wings. The first of two prototypes took to the air for the first time on 9 May 1949. In December 1952 one of them flew faster than the speed of sound. (Bob Snyder)
25. The Fairchild XC-120 Pack-Plane with cargo pod attached. (US Air Force)
26. In common with the British SR.A/1, the Convair Sea Dart was built as an experimental turbojet-powered flying-boat fighter. However, unlike the SR.A/1, as taxiing speed increased the fighter lifted above the water on one or two retractable hydroskis, thereafter aquaplaning until take off. Five Sea Darts were built, the first (XF2Y-1) flying on 9 April 1953.

▼23

24▲

25▲ 26▼

NAVY

27. Convair XFY-1 in vertical flight.
28. The YF-93A, a high-speed penetration fighter produced by North American, was a direct development of the successful F-86 Sabre. Two prototypes were built. Each was powered by a Pratt & Whitney J-48 turbojet with afterburner and featured flush air intakes and twin-wheel main undercarriage units. These were tested at the NACA Flight Test Laboratory at Muroc. (US Air Force)
29. Developed as a successor to the US Navy's piston-engined AD Skyraider, the Douglas A2D Skyshark had the distinction of being the first US combat aeroplane with a turboprop engine. This illustration, dated 11 May 1950 (just over two weeks before the XA2D-1's first flight), clearly shows the co-axial contra-rotating propellers for the Allison T40-A-6 engine.
30. This photograph was released on 27 March 1952, and was the first to show a Skyshark airborne. Engine development problems resulted in only seven Skysharks being built as engine test aircraft. Under the starboard wing can be seen one of three 2,000lb bombs carried by the A2D-1, featuring a special streamline casing.

◀27

28▲

29▲ ▼30

▲31 ▼32

31. The French Sud-Est SE 2410 Grognard is one of the most unusual looking prototypes in this book. Its strangely 'humped' appearance is given by the installation of two Nene turbojet engines one above the other and the use of a top fuselage air intake. The Grognard was designed as an experimental ground-attack aircraft and first flew on 30 April 1950. It ended its life as an armaments test aircraft, as did the refined SE 2415.
32. Something missing? Proving that it could fly with or without its detachable cargo pod was this Fairchild XC-120 Pack-Plane, under test for the USAF. First flown on 11 August 1950, the Pack-Plane was based on the C-119 Packet military transport. No production was undertaken.
33. A front-on view of the XC-120 Pack-Plane. The flat under-surface of the fuselage allowed the detachable pod to be attached directly underneath (see colour illustration 25).

33▼

▲34

34. The Short S.B.3 pictured on the day of its first flight, 12 August 1950. The S.B.3 was developed from the Sturgeon as an anti-submarine aircraft for possible use by the Royal Navy, and featured a bulged forward fuselage which accommodated search radar in a lower radome and two operators in a cabin above.

35. One of the two S.B.3s that were completed from Sturgeon airframes. Note the jetpipe for the port side Armstrong Siddeley Mamba engine. The position of the jetpipes caused stability problems.

▼35

36▲

37▲ 38▼

36. The Avro Ashton Mk.1 (WB490) was built as a large research aircraft to test the operation of turbojet engines at high altitude. This Rolls-Royce Nene-powered aircraft was the first of six Ashtons ordered by the British Ministry of Supply, first flying on 1 September 1950.
37. Ashtons were also used for cabin pressurization and air-conditioning studies, engine and instrument testing and many other purposes. The third Ashton illustrated (WB492) was the first of three Mk. 3s and was used in the development of an advanced radar bomb sight. The ventral radome that carried the sight is visible under the fuselage. The underwing containers carried bombs.
38. A side view of Ashton Mk. 3 WB492. The twin-wheel main undercarriage units can be seen to retract into the paired engine nacelles.

▲39

▲40 ▼41

42▲

39. On 13 April 1949, the Sud-Ouest SO M.2 made its first flight. Powered by a Rolls-Royce Derwent turbojet engine, it was a scaled-down version of the projected SO 4000 bomber and was used to flight test various design innovations prior to the construction of the bomber. One interesting feature of the SO M.2 was its unique undercarriage arrangement.
40. Using data gained from the SO M.1 glider and SO M.2, the SO 4000 high-performance bomber was produced, first flying on 15 March 1951. Powered by two French-built Nene turbojet engines and carrying a crew of two in tandem, bombs were carried in a centre-section bay, above which were the fuel tanks. This bomber remained a prototype, but spawned the SO 4050 Vautour.
41. This illustration shows the unusual undercarriage of the SO 4000, the main units comprising tandem pairs of large wheels.
42. The Breguet 960 Vultur was designed and built in prototype form as a side-by-side two-seat naval strike aircraft. Power was provided by a nose-mounted Armstrong Siddeley Mamba turboprop and an Hispano-Suiza Nene turbojet engine in the tail. A speed of 400km/hr (249mph) was achieved on Mamba power for cruising, and this increased to nearly 900km/hr (559mph) with both engines in use.
43. Only two Vulturs were built, the first flying on 3 August 1951. The second prototype was sent to the Royal Aircraft Establishment at Farnborough, England, where it was used in simulated aircraft carrier operations. The design was abandoned in 1954. Type 960-02 was modified as a test aircraft for the Type 1050 Alizé.

43▼

▼44 ▲45

44. An aircraft that used HMS *Eagle* for carrier trials (in 1953) was the Bristol Type 173 helicopter. Only the company's second helicopter, it was powered in its original Mk. 1 form (XF785) by two Alvis Leonides Mk. 73 radial engines, making it the first British helicopter with more than one engine. The first flight was achieved on 3 January 1952, originally as G-ALBN, prototype for the expected British European Airways production version.
45. XF785 on board HMS *Eagle*, with rotors folded. (Admiralty)

▲46 ▼47

48▲ 49▼

46. After failing to win military orders, the Type 173 remained flying in prototype form for twin-engined, twin-rotor helicopter research. The second of the five (XH379) also flew in BEA colours for handling trials in 1956. XH379 had been used previously for naval trials with four-blade rotors. Here XF785 flies at Farnborough, in 1957, fitted with four-blade rotors and a Mk. 3-type tailplane with endplates.
47. From the Type 173 Bristol developed the Type 192, which eventually entered RAF service as the Belvedere. The first Type 192 was also the first production example, and is illustrated here to allow comparison with XF785 in the preceding photograph.
48. First flown late in 1951, the Swedish Saab-210 Draken was a small 16ft-span research aircraft, built and flown to evaluate the then-new double-delta wing configuration. The data gained from more than 500 flights helped in the development of the Saab-35 Draken combat aircraft. Here, the aircraft has deployed its drag parachute.
49. During the course of testing, the forward part of the Saab-210 was altered. As can be seen from photograph 48, originally the air intakes for the Armstrong Siddeley Adder turbojet were close to the nose. After modification, these were cut back, and this configuration was used for the Saab-35. The pilot, B. R. Olow, is seen here entering the cockpit.

▲50

▲51 ▼52

50. Escorted by a Supermarine Swift, the first of two prototype Supermarine Type 508s (VX133) shows off its unique configuration. This aircraft, first flown on 31 August 1951, was then the most powerful naval fighter in the world designed for use from aircraft carriers. Power was provided by two Rolls-Royce Avon turbojet engines.
51–53. In May 1952 the Type 508 completed trials on board the aircraft carrier HMS *Eagle.* These illustrations show the second Type 508 (VX136) during carrier trials. (Admiralty)

53▲

54. As the next stage in the development of an operational twin-turbojet naval fighter, the Type 508s were superseded by the Type 525 (VX138), here seen on its first flight on 28 April 1954. Also Avon powered, the Type 525 featured swept wings and a conventional swept tail unit. The production Scimitar was developed from the Type 525.
55. The pilot of the Type 525 for its flight on 28 April was Lieutenant-Commander M. J. Lithgow, seen here in his flying suit.

54▲ 55▼

▲56

56. Often referred to as Britain's first four turbojet-engined bomber, the Short S.A.4 Sperrin in fact flew after the prototype Vickers Valiant, the first of two Rolls-Royce Avon-powered S.A.4s initially flying on 10 August 1951. However, as an earlier design, the S.A.4 has its place in British aviation history. The aircraft illustrated is the second prototype S.A.4.

57. Without production orders, the Sperrins became engine and equipment test-beds. In 1955 the first Sperrin flew with three Avons and a large de Havilland Gyron turbojet installed. The following year one of the Avons was replaced by a second Gyron. This photograph clearly shows that the lower portion of the port engine nacelle has been enlarged to accommodate the Gyron.

58. The tandem seating arrangement for pilot and co-pilot in the Boeing YB-52 made the aircraft only just recognizable as a Stratofortress. The progenitor of the famous B-52 series of bombers for the USAF, the YB-52 beat the XB-52 into the air by completing its maiden flight on 15 April 1952.

57▲ 58▼

▲59

59. This diminutive helicopter, known as the 'Jet Jeep' but correctly designated the American Helicopter XH-26, was developed for the US Army. Accommodating only a pilot and powered by two rotor-mounted AJ-7.5-1 pulse-jets, it had been designed to be air-dropped in a container and assembled in the field by two men in about twenty minutes. Its roles could include observation, reconnaissance and stretcher carrying. The maximum speed of the aircraft was about 129km/hr (80mph). The first 'Jet Jeep' initially flew on 30 June 1952.
60. Although a total of five XH-26 helicopters was ordered for service evaluation, the programme was subsequently abandoned. This illustration shows four 'Jet Jeeps' under construction.
61. Just lifting from the ground is the Hughes XH-17, a giant flying-crane helicopter with a 130ft-diameter rotor. First flown on 23 October 1952, the XH-17 was powered by two modified General Electric turbojet engines which fed gas pressure to rotor-tip outlets. Built to a USAF contract, its maximum take-off weight was 19,504kg (43,001lb).
62. Martin SeaMaster.
63. North American F-107A.

▲60 ▼61

62▲ 63▼

▲64 ▼65

66▲

64. The Italian Aerfer Ariete was an experimental single-seat lightweight fighter. It was powered by a Rolls-Royce Derwent turbojet engine carried in the nose and exhausting under the fuselage, and a Rolls-Royce Soar turbojet in the tail. Air for the latter came via a retractable intake in the upper fuselage. The Soar was operated for periods of high-speed flight only. First flown on 27 March 1958, the supersonic Ariete remained a prototype.
65. Using some component parts from the Sikorsky S-56, Westland produced its Westminster. Two versions were envisaged for military and civil use, one as a transport for up to 43 troops/passengers, stretchers or freight, and the other as an open-structure flying crane. It was in the latter form that the Westminster first flew, on 15 June 1958, powered by two Napier Eland turboshaft engines. However, the Westminster was abandoned in 1960.

66. On 2 December 1952, the Short S.B.5 research aircraft flew for the first time. It had been built to investigate the flight characteristics of swept wings at low speed, and for this the angle of sweep could be changed on the ground, the tailplane moved from T-tail to conventional position and its angle of incidence varied. This illustration shows the S.B.5 with a high tailplane and a 50° sweep to the wings.
67. As testing progressed, the S.B.5 flight-tested wing sweepback angles of 60° and 69°, with a high and low-mounted tailplane and a low tailplane respectively. The data from these flights was used in the design of the English Electric P.1 experimental supersonic fighter. Here the S.B.5 has a wing sweep of 60° and a low-mounted tailplane.

67▼

▲68 ▼69

68. France was one of a few countries that believed there was a future for mixed-power interceptors. In a bid to develop an operational interceptor, Sud-Ouest first constructed a research aircraft known as the SO 9000 Trident I. This first flew on 2 March 1953, with only its wingtip-mounted Turboméca Marboré II turbojets in use.
69, 70. The next stage was to install an SEPR.481 rocket motor in the tail of the Trident I and, on 4 September 1954, the aircraft flew with the rocket motor ignited.
71. Confident in the design, Sud-Ouest produced two new prototypes as SO 9050 Trident IIs. The first of these initially flew on 17 July 1955, using only its Dassault Viper turbojets. Rocket power was first employed on 21 December 1955. On 3 May 1957, the first of six pre-production Trident II interceptors flew, but the programme was abandoned soon after. Almost a year later, on 2 May 1958, one pre-production Trident II with Turboméca Gabizo turbojets set an altitude record of 24,217m (79,450ft). The maximum speed achieved by the Trident II was Mach 1.95. This Trident II is seen blasting away on rocket power.
72. A Trident II on turbojet power, carrying a Matra air-to-air missile under its fuselage.

70▲

71▲ 72▼

▲73 ▼74

75▲ 76▼

73. In April 1949, the French Leduc 0.10 made its first powered flight. This was the world's first manned aircraft to use a ramjet engine. It had to be air launched, as the engine would not operate until it received high-velocity air that passed through the double-skinned tubular fuselage. Developed from the 0.10 were the 0.16 and 0.21, the latter as a further stage toward the 0.22 interceptor prototype. Two 0.21s were built. The one illustrated is being readied for flight on top of the Languedoc 'motherplane'.
74. An 0.21 in flight, after release from the Languedoc. The first flight was achieved on 16 May 1953.
75. The glazed nose of the 0.21 slid forward to allow entry into the cabin, which was sealed by an inflatable ring when closed.
76. The pilot of the 0.21 sat in a jettisonable glazed cabin forward of the annular air duct for the ramjet engine.

▲77 ▼78

79▲

77. The French Potez 75 was designed as a cheap armoured two-seat ground support, anti-tank and observation aircraft. It was powered by a Potez 8-D.32 piston engine mounted in the rear of the nacelle and could carry a cannon in the nose and bombs, rockets or missiles underwing. In refined form it was ordered for the French Army, but in 1957 the contract was withdrawn.
78. Another French project of the early 1950s involved the development of a single-seat tactical fighter that could operate from virtually any fairly firm and flat surface, including beaches. Known in prototype form as the Sud-Est SE 5000 Baroudeur, it was turbojet powered and could carry two 1,000kg bombs.
79. The secret of the Baroudeur's success lay in its undercarriage arrangement. In place of the usual wheels, it could take off and land on retractable skids with replaceable bottom edges. Here, one of the three SE 5003 pre-production Baroudeurs built demonstrates a skid landing.
80. An alternative form of take off used by the Baroudeur involved the use of a rocket-powered trolley with low-pressure tyres. The trolley, which was reusable, remained on the ground and was fitted with brakes.

80▼

▲81

81. The McDonnell XV-1 was an experimental converti-plane, developed in close co-operation with the USAF and US Army. It could be used as a four-seater or accommodate a pilot, two stretchers and attendant. The XV-1 used a pressure jet-driven rotor for vertical flight and a piston engine-driven pusher propeller, wings and autorotating rotor for horizontal flight. It first flew on 11 February 1954, but the programme was abandoned in 1957.

82. The name Cessna is not one that is readily associated with the manufacture of helicopters. However, in 1953 the company flew its first test helicopter from which the YH-41 Seneca was eventually developed for evaluation by the US Army. It was a four-seater with its engine in the nose. The first prototype for evaluation is seen here in 1957, in the hands of First Lieutenant A. E. Lush.

82▲

83. This peculiar looking aircraft was a standard Gloster Meteor Mk. 8, modified by Armstrong Whitworth as a test aircraft to evaluate the semi-prone pilot's position for high-performance aircraft. As can be seen, the nose was extended to accommodate a new cockpit fitted with full flying controls and a semi-prone position.

83▼

▲84 ▼85

84. WG760, seen here, was one of two flying English Electric P.1A prototype fighters. The first made its maiden flight on 4 August 1954. Both prototypes were each powered by two Bristol Siddeley Sapphire turbojet engines. The P.1A was the first British aircraft designed for supersonic flight. The Lightning interceptor was developed from the P.1As and P.1Bs.

85. Exactly 28 months after the first flight of a P.1A, the first of three P.1Bs took to the air. Basically prototypes of the operational Lightning F.Mk 1, the P.1Bs could be armed with two Firestreak missiles in addition to two 30mm Aden cannon. Power was provided by Rolls-Royce Avon RA.24R turbojets. A recognition feature of the P.1B was its new air intake with a radar-carrying centre-body.

86. Prior to the construction of the XFV-1 (see illustration **1**), a quarter-scale model was tested at NACA's Ames Aeronautical Laboratory. The tethered model was remotely controlled and powered through a cable.

87. Convair's tail-sitting VTOL fighter was the XFY-1 Pogo. The wing and tail configurations differed considerably from those of the Vertical Riser. Prior to free flights, the Pogo carried out many tethered 'flights' from a 195ft high rig. Here, the pilot, J. F. 'Skeets' Coleman, occupies his pivoting seat.

86▲ 87▼

NAVY
NAVY

UNITED STATES NAVY
CONVAIR
U.S. NAVY
J4

88. Convair built this mobile housing and maintenance shed, seen here with the Pogo about to be moved.
89, 90. A mobile trolley was developed for raising and lowering the Pogo. Once in the vertical position, the pilot entered the cockpit via a 20ft ladder. Interestingly, in the background Tradewind flying-boats can be seen under construction, their huge tails protruding from the hangar.
91. On 2 August 1954, the Pogo made its first vertical 'up and down' flight, making 'Skeets' Coleman the first pilot of a VTOL fighter. Seventy similar flights were followed by the first transition to horizontal flight and back, on 2 November. However, the Pogo also remained a prototype. The X marker indicates the 50 sq ft of area on the long Lindbergh Field runway needed for take-off and landing.

90▲ 91▼

▲92

▲93 ▼94

92. The Hiller HJ-1 Hornet was a small two-seat helicopter powered by two rotortip-mounted Hiller 8RJ2B ramjet engines. Over the course of several years, twelve YH-32 and three HOE-1 Hornets were built for and evaluated by the US Army and Navy respectively. The original Hornet prototype flew in 1950 and featured a rudder instead of a tail rotor, as seen here during early demonstrations. Note the open structure and the stretcher patient.
93. The Hornet was revised several times to comply with military demands. This Army YH-32 has a tail rotor but is without the rear portion of the boom that carried tail surfaces with marked anhedral.
94. On 23 October 1953, the first US Piasecki PV-15 Transporter took to the air. Built to a USAF contract, this 40-passenger or freight-carrying twin-rotor helicopter was powered by two Pratt & Whitney piston engines. It carried the military designation YH-16. The second Transporter, with more-powerful Allison T38 turboshafts, became the YH-16A. Neither version entered production. The example illustrated is the YH-16, seen over Philadelphia during its first flight.
95. The Hiller 'Flying Platform', first flown in January 1955, is seen here in Navy markings. It comprised co-axial contra-rotating fans enclosed in a ring and powered by two small piston engines. A further, much refined prototype appeared in 1956 and became the US Army's experimental VZ-1E Pawnee. Each platform was guided by the pilot leaning in the direction he wished to travel (see illustration **2**).
96. The British two-seat Fairey Ultra Light Helicopter was designed and built primarily as a military observation helicopter, although other military and civil roles were within its capability, including the transportation of slung cargo. The first of the two prototypes ordered for Army evaluation initially flew on 14 August 1955. Like so many other experimental helicopters of the period, it used rotortip-mounted pressure-jets, fed with compressed air from a Blackburn-built Palouste 500 air generator. Cruising speed was an excellent 158km/hr (98mph).

95▲ 96▼

▲97

97. Another Meteor Mk. 8 conversion (see illustration **83**) was VZ517, modified to flight test an Armstrong Siddeley Screamer rocket motor carried under the fuselage. Flight clearance was given in December 1955.

98. Although a total of two XP6M-1 prototypes, six YP6M-1 pre-production and three P6M-2 production aircraft was built, the Martin 275 SeaMaster did not achieve full operational use and so has a place here. An incredible patrol flying-boat, well equipped, well armed, and powered by four Allison or Pratt & Whitney turbojet engines mounted over the marked-anhedral wings, it could demonstrate a maximum speed of 966km/hr (600mph). Furthe production aircraft were cancelled. Illustrated is an XP6M-1, the first of which flew on 14 July 1955.

99. A Martin P6M-2 SeaMaster takes off.

▼98

99

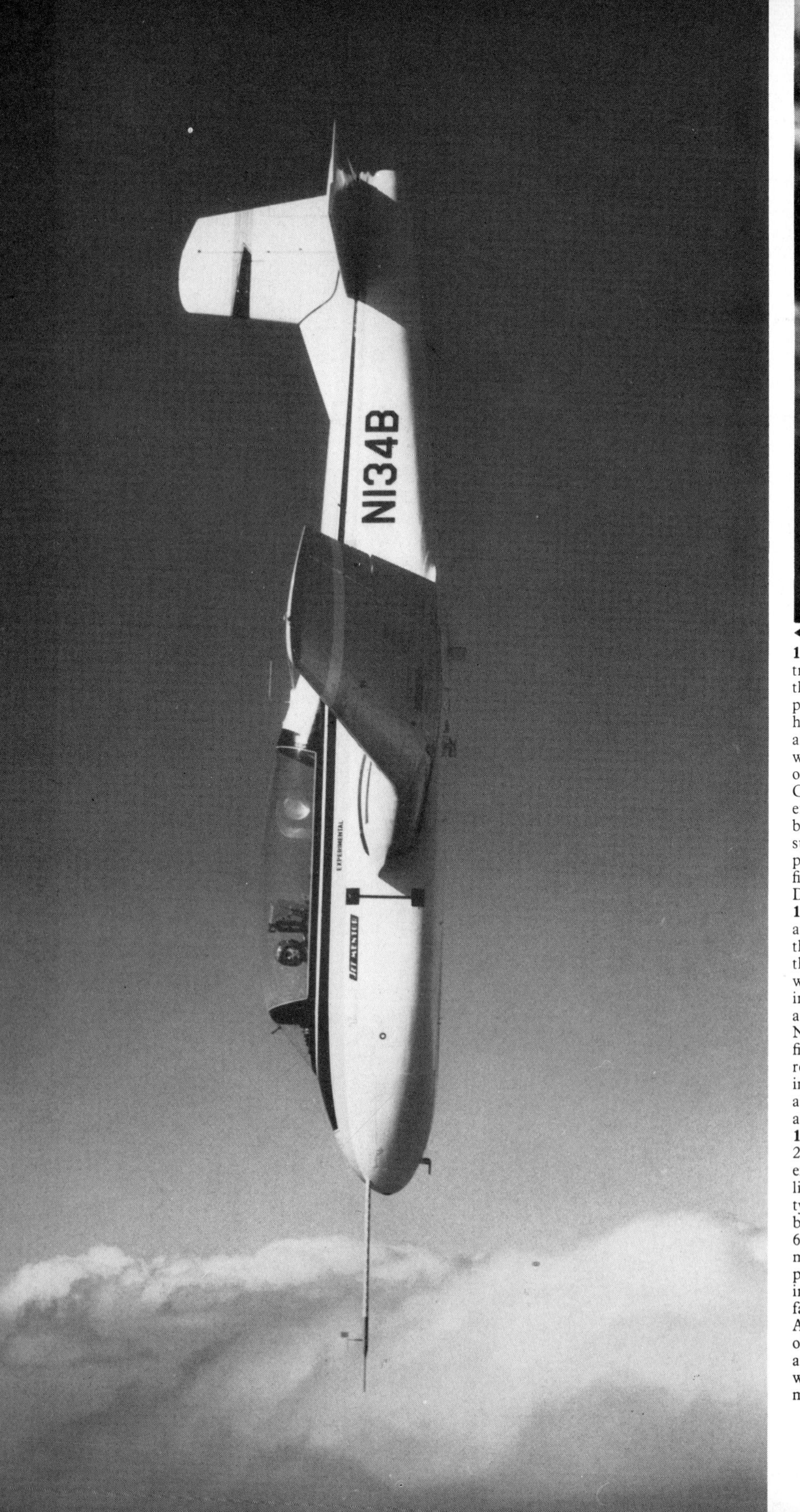

◀**100**

100. To fulfil the role of jet trainer, Beechcraft produced the Model 73 Jet Mentor as a private venture. Low cost but high performance were the aims and, indeed, the aircraft was capable of 463km/hr (288mph on the power of its single Continental J69-T9 turbojet engine. The airframe was based upon Beechcraft's successful T-34 Mentor piston-engined trainer. The first flight was achieved on 18 December 1955.

101. The six pusher engines and four turbojets identify this aircraft as a member of the Convair B-36 family, but with a difference. First flying in 1955 but officially announced in early 1956, the NB-36H Crusader became the first aircraft to carry a nuclear reactor. This was used to investigate the problems associated with shielding aircraft from radiation.

102. The Sud-Aviation SE 212 Durandal was an experimental single-seat lightweight interceptor, typically French by having both Atar turbojet and SEPR 65 rocket power. The rocket motor and associated liquid propellant tanks were carried in a detachable under-fuselage fairing. First flown on 20 April 1956, it proved capable of Mach 1.5. The main armament of the Durandal was a Matra 510 air-to-air missile.

101▲ 102▼

▲103 ▼104

105▲

103. The success of the North American F-100 Super Sabre led to the development of an experimental Mach 2 fighter-bomber derivative, which became the F-107A. Three prototypes were constructed, the first flying on 10 September 1956. Each featured special, large bifurcated air intakes above the fuselage designed to slow the velocity of air to the engine at supersonic speeds.
104. Originally designated F-100B, the F-107A also featured an all-moving fin and spoiler-type ailerons. The streamline fairing under the fuselage enclosed a large bomb.

105. The French Breguet 1001 Taon was designed and built to a NATO specification for a single-seat lightweight tactical strike fighter (as was the later operational Italian G.91). Three prototypes were built, the first making its maiden flight on 26 July 1957. Each was powered by a single Bristol Orpheus turbojet engine.
106. Despite having been judged the best submission at design stage, the Taon did not enter production. (Photo AMD-BA)

106▼

▲107 ▼110
▲108 ▼111

109▲

107–112. This sequence of photographs shows the second of two Short SC.1 VTOL research aircraft (XG905) during a full transition from vertical to horizontal flight. The SC.1 was the first British VTOL aeroplane. It used four Rolls-Royce RB.108 turbojet engines for vertical lift and one installed horizontally for main forward thrust. The first flight, on 2 April 1957, was performed conventionally and it was not until 25 October 1958 that a free vertical flight was achieved by XG905. Both SC.1s flew during much of the 1960s.

112▼

▲113 ▼115

114▲

113. The Saunders-Roe S-R.53 was the first piloted aircraft of British origin to use mixed power. Two prototypes were built, the first initially flying on 16 May 1957. The jetpipe for the Armstrong Siddeley Viper turbojet engine was fitted just below the T-tail, under which (and seen here in use) was located the de Havilland Spectre rocket motor. More advanced prototypes developed from the Mach 2.4-plus S-R.53s were being built when the programme was officially dropped.
114. On 25 March 1958, Avro Aircraft of Canada flew the first of five prototypes of its beautiful CF-105 Arrow interceptor. Canada's first aircraft capable of supersonic speeds in level flight, each was powered by two Pratt & Whitney J75 turbojet engines and carried Sparrow II air-to-air missiles in a large under-fuselage weapons bay. Here, an Arrow 1 prototype leaves the flight test hangar to start engine test runs.
115. During evaluation the second Arrow 1 achieved 1,931km/hr (1,200mph) in level flight. From the Mk. 1 was to have been built the Orenda-powered Mk. 2 production interceptor, 32 of which had been specified initially. However, in February 1959 the Arrow programme was cancelled.

▲116 ▼117

116. An important experimental transport aircraft from France was the Breguet Type 940 Integral, first flown on 21 May 1958. Its main feature was the 'blown wing' system, in which the slipstream from the four turboshaft-driven propellers was directed to the large double flaps to produce lift for STOL operations.
117. The Integral was built as an experimental transport but still had an ample cabin with a rear loading ramp. From the Integral was developed the larger Type 941, capable of accommodating nearly 50 passengers or freight. This flew as a prototype in the 1960s. Here the Integral is viewed prior to take-off.
118. The French Sud-Aviation SE 116 Voltigeur first flew on 5 June 1958, on the power of two Wright radial engines. It was then designated a combat-area support type, capable of performing attack, reconnaissance, casualty evacuation (two stretchers or five passengers) and many other roles.
119. The projected production version of the Voltigeur was the SE 117, normally fitted with two Turboméca Bastan turboprop engines. The third prototype is illustrated here.
120. In 1957 one of the US Navy's WV-2 Warning Star early-warning aircraft was modified into the WV-2E, as illustrated. This was a specialist early-warning radar intelligence aircraft, with a very large radar scanner in a 37ft elliptical cross section radome mounted on the fuselage. However, this was the only conversion, as Grumman aircraft fulfilled this role for the Navy.
121. (overleaf) The French Morane-Saulnier MS.1500 Epervier was a low-cost, multi-purpose combat aircraft, powered by a turboprop engine. First flown on 12 May 1958, it could carry a 450kg (992lb) weapons load when flown as a two seater or a 550kg (1,213lb) load as a single seater. Its cruising speed was a mere 365km/hr (227mph) at an all-up-weight of 2,450kg (5,400lb).

118▲

119▲ 120▼

M S
500

Epervier

▲122

122. On 12 October 1958, Piasecki first flew its revolutionary Model 59-K Sky-Car, known to the US Army as the VZ-8P Airgeep I. Originally powered by two Lycoming piston engines, it was a four-seat VTOL aircraft for low-level observation and liaison duties, the seats arranged between the two large rotor ducts. As illustrated, the Airgeep was later given Turboméca Artouste IIB turboshaft engines. It was first flown in this form in June 1959.

123. Under contract from the Bureau of Naval Weapons, the Airgeep I was later equipped with floats for test flights from and over water. In this form it was known as the Seageep.

124. From the Airgeep I was developed the more advanced VZ-8P (B) Airgeep II, powered by two Artouste IIC turboshaft engines. This aircraft also featured powered wheels, allowing it to be driven over land when desirable, so making it a true air jeep. First flown in 1962, this version also remained a prototype.

123▲ 124▼

▲125

▲126 ▼127

125. A trial with the Sikorsky S-60 demonstrated the helicopter's ability as a minesweeper, with a specially-equipped pod attached below the boom. As can be seen, the co-pilot sat on a swivelling seat, enabling him to face rearward to observe the cargo. He had a second set of controls.

126. The US Sikorsky S-60 was a large flying-crane helicopter. It was heavily based on the successful S-56, common components including the engines, rotor heads and blades, and drive shafts. The engines, rotor, etc, were mounted on a shallow boom, leaving the area below and to the rear of the pilots' cabin clear for attaching external cargo. During August and September 1959 the US Army evaluated the S-60 carrying a 20-passenger detachable pod, as seen here during a trial with combat engineer troops from Fort Belvoir, Virginia.

127. The final prototype in this book is another from France. The Sud-Aviation SA.3200 Frelon was designed as a triple Turboméca Turmo III-engined transport, air/sea-rescue, minesweeping and armed helicopter. In the first of these roles it could carry up to 24 troops, 15 stretchers or freight. First flown on 10 June 1959, the SA.3200 was developed into the SA.321 Super Frelon production helicopter.